QUANTUM PHYSICS

TABLE OF CONTENTS

INTRODUCTION

Physics utilizes the scientific method to discover the fundamental laws that control light and matter, as well as the effects of those laws. It is presumptively true that there are laws that govern how the world works, and that these laws are at least partly understandable to humans. It is also widely assumed that if full knowledge about the current state of all light and matter were available, those laws could be used to predict anything about the universe's future.

Anything with mass and volume is usually referred to as matter. Many of the theories and laws that explain matter and motion are central to the study of classical physics. For example, the law of conservation of mass states that no mass can be created or destroyed. As a result, when formulating theories to describe natural phenomena, further experiments and calculations of physics take this rule into account. Physics seeks to explain how everything around us works, from the movement of tiny charged particles to the movement of humans, vehicles, and spaceships. In reality, the laws of physics can accurately explain almost anything around you. Consider a smartphone: physics explains how electricity interacts with the device's various circuits. This knowledge aids engineers in choosing the right materials and circuit layout for the smartphone. Consider a GPS; physics defines the relationship between an object's speed, its travel distance, and the time it takes to travel that distance. When you use a GPS in your car, it uses these physics equations to find out how long it would take you to get from one spot to another. Physics will make a major contribution to society through developments in new technology resulting from theoretical breakthroughs.

Chapter 1: UNDERSTANDING QUANTUM PCHYSIC

What is quantum physics, and how does it work? Simply put, physics is the best explanation we have of the origin of the particles that make up matter and the forces by which they interact, and it describes how everything works.

Quantum science explains how atoms work, as well as how chemistry and biology work. At some stage, you, me, and the gatepost are all dancing to the quantum theme. Quantum physics is needed to understand how electrons travel through a computer chip, how photons of light are converted to electrical current in a solar panel or amplify themselves in a laser, or even how the sun continues to burn.

Here is where the challenge – and, for physicists, the fun – begins. To begin with, no single quantum theory exists. There's quantum mechanics, the fundamental mathematical framework that underpins everything, established by Niels Bohr, Werner Heisenberg, Erwin Schrödinger, and others in the 1920s. It describes basic things like the shift in position or momentum of a single particle or a group of a few particles over time.

Quantum mechanics must be combined with other elements of physics, most notably Albert Einstein's special theory of relativity, which explains what happens when things move very quickly, to construct quantum field theories, which explain how things function in the real world.

Three separate quantum field theories are used to describe three of the four fundamental forces that interfere with matter: electromagnetism, which explains how atoms stay together; the strong nuclear force, which explains the structure of the nucleus at the center of the atom; and the weak nuclear force, which explains why certain atoms decay radioactively.

All three theories have been bundled together in a shambles known as the "normal paradigm" of particle physics for the last five decades or so. Although it appears to be bound together with sticky tape, this model is the most thoroughly tested depiction of matter's basic workings ever invented. Its crowning achievement came in 2012 with the discovery of the Higgs boson, the particle that gives all other fundamental particles their mass and whose presence had been predicted as far back as 1964 using quantum field theories.

The effects of experiments at high-energy particle smashers, such as CERN's Massive Hadron Collider, where the Higgs was discovered, which probe matter at its smallest scales, are well explained by traditional quantum field theories. Things get much more complicated if you try to understand how things operate in many less complex cases, such as how electrons travel or don't move through a solid material to make it a metal, an insulator, or a semiconductor, for example. The billions of experiences in these crowded environments necessitate the creation of "powerful field theories" that gloss over some of the gruesome information. Many important questions in solid-state physics remain unanswered due to the complexity of building certain theories, such as why certain materials are superconductors at low temperatures, allowing current to flow without resistance, and why we can't get this trick to function at room temperature.

But there's a big quantum mystery lurking under all of these practical issues. Quantum theory predicts very peculiar things about how matter functions at a fundamental level that is entirely at odds with how things seem to work in the real world. Quantum particles can behave like particles when they are in a single location, or they can behave like waves when they are dispersed across space or in several locations at the same time. How they appear seems to be determined by how we measure them, and before we measure them, they appear to have no definite properties at all, posing a fundamental conundrum about the existence of basic truth.

SIX THINGS EVERYONE SHOULD KNOW ABOUT QUANTUM PHYSICS

Quantum physics is also daunting right from the outset. Though physicists who deal with it daily find it strange and counter-intuitive. It is not, however, nonsensical. There are six main principles in quantum physics that you should consider if you're reading anything about it. If you do so, you'll find quantum mechanics far easier to grasp.

Waves make up everything, even particles.

There are several ways to begin a discussion like this, and this is one of the best: anything in the universe has both a particle and a wave existence at the same time. "Everything is waves, with nothing waving, over no space at all," says a character in Greg Bear's fantasy ideology (The Infinity Concerto and The Serpent Mage), explaining the basics of sorcery. That's always struck me as a romantic definition of quantum physics — everything in the universe has a wave nature at its heart.

Of course, everything in the universe is made up of particles. This seems to be insane, but it is an experimental reality that was discovered through a remarkably familiar process:

Of course, representing physical objects as particles and waves is inherently a little sloppy. Quantum physics defines a third group of artifacts that share some properties of waves (a characteristic frequency and wavelength, some scattered through space) and some properties of particles (they're usually countable and can be localized to some extent). This has sparked a vigorous debate in the physics education community about whether it's acceptable to speak about light as a particle in introductory physics courses; not because there's some disagreement about whether the light has a particle nature, but because calling photons "particles" rather than "excitations of a quantum field" may lead to some student misunderstandings. I don't agree with this because labeling electrons "particles" raises many of the same questions, but it does provide a reliable source of blog discussions.

The often perplexing terminology used by physicists to discuss quantum phenomena illustrates this "door number three" existence of quantum artifacts. The Higgs boson was discovered as a particle at the Large Hadron Collider, but physicists often refer to the "Higgs field" as a delocalized entity that fills all of space. This occurs because it is more convenient to analyze Higgs field excitations in terms of particle-like properties in certain situations, such as collider experiments, whereas it is more convenient to discuss the physics in terms of interactions with a universe-filling quantum field in other circumstances, such as general discussions of why certain particles have mass. It's just a different way of representing the same mathematical object in a different language.

Quantum Physics Is Discrete

The word "quantum" comes from the Latin for "how many," and it refers to the fact that quantum models often contain something that comes in discrete quantities. Quantum fields contain energy that is integer multiples of some fundamental energy. This is related to the frequency and wavelength of light, with high-frequency, short-wavelength light having a broad characteristic energy and low-frequency, long-wavelength light having a small characteristic energy. However, in all cases, the total energy present in a light field is an integer multiple of that energy— 1, 2, 14, 137 times— never a strange fraction like one-and-a-half or the square root of two. This property can also be seen in atoms' distinct energy levels and solids' energy bands, where some energy values are permitted while others are not. Because of the discrete nature of quantum physics, atomic clocks operate by using the frequency of light associated with a transition between two permitted states in cesium to hold time at a level that does not necessitate the much-discussed "leap second" introduced last week.

Ultra-precise spectroscopy can also be used to search for dark matter, which is one of the reasons for the establishment of a low-energy fundamental physics institute.

Also, fundamentally quantum phenomena like black-body radiation tend to include continuous distributions, which isn't always clear. However, digging into the mathematics shows a granularity to the underlying truth, which is a large part of what contributes to the theory's strangeness.

Quantum Physics Is Probabilistic

One of the most shocking and (at least historically) divisive aspects of quantum physics is that the result of a single experiment on a quantum system cannot be predicted with certainty. When physicists predict the outcome of an experiment, they always give a probability for discovering each of the specific possible outcomes, and comparisons between theory and experiment always include inferring probability distributions from a large number of repeated experiments.

A "wavefunction" is a mathematical representation of a quantum system, which is usually represented in equations by the Greek letter psi: There's a lot of debate on what this wavefunction represents, and it's divided into two camps: those who believe the wavefunction is a real physical thing (jargon term: "ontic" theories), and those who believe the wavefunction is merely an expression of our knowledge (or lack thereof) about the universe ("epistemic" theories).

The probability of finding an outcome is not directly given by the wavefunction in either class of foundational model, but by the square of the wavefunction (loosely speaking; the wavefunction is a complex mathematical object (meaning it involves imaginary numbers like the square root of negative one), and the operation to get probability is slightly more involved, but "square of the wavefunction"). The "Born Law," named after German physicist Max Born, who proposed it (in a footnote to a paper in 1926), is seen by some as an unattractive afterthought. Some parts of the quantum foundations group are working hard to find a way to derive the Born rule from a more fundamental principle; so far, none of them have been completely successful, but it's generating a lot of interesting research.

This is also the feature of the theory that allows particles to exist in several states at the same time. Only probability can be predicted, and before a calculation that decides a specific outcome, the system being evaluated is in an indeterminate state that mathematically maps to a superposition of all possible outcomes of varying probabilities. If you think of this as the system being in all of the states at once or only one unknown state depends a lot on how you feel about ontic versus epistemic models, but all are constrained by the next point on the list:

Quantum physics is a nonlocal science.

The last major contribution Einstein made to physics was underappreciated, largely because he was incorrect. Einstein made a simple mathematical statement of something that had been troubling him for some time, a concept that we now call "entanglement," in a 1935 paper with his younger colleagues Boris Podolsky and Nathan Rosen (the "EPR paper"). According to the EPR paper, quantum physics permitted the existence of systems in which measurements taken at widely separated locations could be correlated in such a way that the outcome of one was decided by the outcome of the other. They claimed that this meant that the measurement results had to be predetermined by some common factor since the alternative would entail sending the result of one measurement to the position of the other at speeds greater than the speed of light. Thus, quantum mechanics must be incomplete, a mere approximation to a deeper theory (a "local hidden variable" theory, in which the results of a particular measurement are determined by a factor common to both systems in an entangled pair (the "hidden variable"), rather than something further away from the measurement location than a signal might travel at the speed of light ("local").

For around thirty years, this was considered to be a curious footnote since there appeared to be no way to test it, but in the mid-1960s, Irish physicist John Bell found out the ramifications of the EPR paper in greater detail. Bell demonstrated that quantum mechanics can predict stronger correlations between distant measurements than any other conceivable theory of the kind favored by E, P, and R. This was tested experimentally by John Clauser in the mid-1970s, and Alain Aspect's early 1980s experiments are generally regarded as conclusively demonstrating that these intertwined structures cannot be explained by any local hidden variable theory.

The most common explanation for this finding is that quantum mechanics is non-local, meaning that the results of measurements taken at a specific location can be affected by the properties of distant objects in ways that cannot be clarified using light-speed signals. This does not, however, allow for the transmission of information at speeds faster than the speed of light, despite several attempts to do so using quantum non-locality. Refuting these has proven to be a remarkably fruitful endeavor; for more information, see David Kaiser's How the Hippies Saved Physics. Quantum non-locality is also at the heart of the knowledge issue in evaporating black holes, as well as the "firewall" debate that has sparked a lot of recent debate. There are even some radical ideas involving a mathematical relation between entangled particles and wormholes, as defined in the EPR paper.

The Scale of Quantum Physics Is (Mostly) Miniscule

Quantum science has a reputation for being strange because its predictions are so dissimilar to our daily reality (at least for humans; the conceit of my book is that it isn't as strange to dogs). This occurs because the effects involved become smaller as objects become larger— if you want to see unmistakably quantum behavior, you simply want to see particles behave like waves, and the wavelength decreases as momentum increases. The wavelength of a macroscopic object such as a dog walking through the room is so small that if you extended it to the scale of the entire Solar System, the dog's wavelength will be around the size of a single atom inside that solar system.

This means that quantum phenomena are often restricted to the size of atoms and fundamental particles, where masses and velocities are small enough for wavelengths to become large enough to detect directly. However, there is a concerted effort in a variety of areas to increase the size of devices that display quantum effects. I've written a lot about Markus Arndt's experiments showing wave-like activity in larger and larger molecules, and there are a lot of groups working on using light to slow the motion of chunks of silicon down to the point that the distinct quantum nature of the motion becomes apparent. There have also been reports that this could be done with suspended mirrors weighing several grams, which would be incredibly cool.

Chapter 2: HISTORY OF QUANTUM TECHNOLOGY

The quantum revolution is on its way, and it will take some major leaps of logic from some of the greatest minds of the twentieth century to get us there.

Popular names like Einstein, Bohr, Pauli, Heisenberg, Schrodinger, and others have forever changed the way we think about the universe in the history of the quantum revolution.

Many physicists' smug self-satisfaction at the start of the twentieth century that "there is nothing to be discovered in physics now; all that remains is more and more precise measurement" (attributed to Lord Kelvin, 1900) quickly gave way to a staggering set of advances that swept away comfortable ideas and replaced them with theories that, while proving to be extremely useful, strained physicists' minds.

The outstanding lineup of speakers at the Belgian physics conference in 1911. Victor Goldschmidt, Max Planck, Rubens, Somerfeld, Lindemann, Louis Victor De Broglie, Knudsen, Hasenohrl, Hostelet, Herzen, James Hopwood Jeans, Ernest Rutherford, Heike Kamerlingh-Onnes, Albert Einstein, Paul Langevin, Victor Goldschmidt, Max Planck, Rubens, Somerfeld, Lindemann, Louis Victor De Broglie, Knud Walther Nernst, Marcel Louis Brillouin, Ernest Solvay, Hendrik Lorentz, Otto Heinrich Warburg, Jean Baptiste Perrin, Wilhelm Wien, Madame Marie Curie, and Jules Henri Poincare are seated at the table from left to right. Photo credit: Getty Images and a warm welcome to Jason Isaacs

"BE STILL AND CALCULATE"

Max Planck, a German theoretical physicist, and later Albert Einstein demonstrated that light is divided into packets, the size of which is determined by the frequency of vibration. Following that was Frenchman Louis de Broglie, who demonstrated that not only does light behave like a particle, but particles can also behave like light – and that they, too, can behave like waves. Another German, Werner Heisenberg, discovered that it was impossible to accurately calculate both the location and velocity of a particle at the same time.

Too many certainty were broken in the first half of the twentieth century.

Quantum leap in computer simulation

The belief that what we measure is true, that particles and waves are fundamentally different, that time is the same for all observers, that nature is fundamentally predictable, and that with enough experience, we can predict how processes will evolve with certainty has been replaced by a view of reality in which the best we can do is predict probabilities. Many scientists, however, had come to terms with the weird new way of looking at the universe by the middle of the twentieth century because quantum mechanics was so damned useful.

Students who protested, "But sir, it doesn't make sense that a particle can be in two places at the same time," we're told to "shut up and measure," because quantum mechanics' ability to reliably predict behavior was astounding.

New quantum-physics-based technology, such as semiconductors and computers, lasers and communications, the internet and GPS, as well as MRI and PET imaging, arose to the delight of a public mostly ignorant of and unconcerned with the fact that even many physicists are unfamiliar with the fundamental understanding of quantum mechanics.

So, what are the crucial moments in the history of quantum physics that inspired the growth of quantum computing and quantum technology?

THE ENTANGLEMENT CONUNDRUM

Well, the first arose from a desire to comprehend the essence of truth.

Let's say we have two particles, A and B, that have emerged from the breakup of a spinning molecule and are spinning in the same direction, either up or down, but we don't know which. These particles are now known as entangled particles. Let's assume I calculate one particle and discover it's spinning. I know that if I weigh the second particle, I'll discover that it's spinning up as well, and vice versa. We know from Heisenberg that measuring a particle changes it – but does this mean that measuring A changes the state of B as well?

Now, Einstein famously referred to this as "spooky activity from afar," and he didn't believe it was possible.

The other possibility is that A and B hold a secret coded code that tells the particles whether they are spinning up or down, even though we don't know what it is.

We had no way of knowing the answer in the 1930s, and the argument seemed as futile as wondering whether a tree falls in the forest and no one hears it. This seemed to be more philosophical than physics.

However, in 1964, John Bell, a Northern Irish physicist, suggested an experiment to distinguish between these two possibilities. In 1981, Frenchman Alain Aspect used Bell's ideas to conduct experiments on pairs of entangled photons, revealing unequivocal results: if two particles are entangled, measuring one of them without disrupting the other is impossible.

Niels Bohr, a physicist, won the Nobel Prize in Physics in 1922 for his work on the structure of atoms. Photo credit: Getty Images

If this is the case, it means that B is aware that A has been measured although A and B are separated by large distances – so large that no signals could travel from A to B during the measurement time. Of course, this means that the optimal signal propagation speed is the speed of light.

And, as difficult as it is to accept using common sense, that is how the world works.

Some may consider this a flaw, but for quantum computers, it is a critical function. The read-out or measurement phase in a traditional machine is simply a way of peering into the computer and getting the result.

Less 'beam me up,' more 007 with photon teleportation

The read-out does not affect the calculation and does not add to it. If I read out the state of one of the qubits in a quantum computer, not only does the state of that qubit change, but all the qubits that are entangled with the qubit that I have read out change as well.

As a result, the measurement becomes a crucial component of the calculation.

I might not even look at the outcome of the qubit I read out because my goal is to alter the state of the other qubits, not to know what the result is. This is a whole new way of thinking, and quantum algorithms make good use of it.

It applies the old philosophical chestnut of the "falling tree in the lonely forest" to technical applications, much as the controversy about the origin of light (is it a wave or a particle?) in the early days of quantum mechanics found practical application in lasers and computer chips.

UP AND ATOM

A talk was the second defining quantum event.

Richard Feynman, a physicist and mathematician, was awarded the Nobel Prize in Physics in 1965. Photo credit: Getty Images

Richard Feynman, a theoretical physicist, posed a seemingly simple question in 1959: could we manipulate matter on the scale of individual atoms to produce nanoscale devices, computers, and new chemicals simply by arranging the atoms in the desired order?

Unsurprisingly, in an era when no one had ever imaged a single atom, let alone manipulated one or arranged atoms in a pattern on a surface, the talk drew little attention. In 1981, Germans Gerd Binnig and Heinrich Rohrer used a scanning tunneling microscope to photograph single atoms, changing everything.

For the first time, we were able to see atoms rather than just guess their presence.

A big discovery in a tiny package

In 1989, IBM scientists used 35 xenon atoms on a nickel surface to spell out the company's name.

Then, in 2013, they released A Boy and His Atom, the world's smallest film, which is a stop-motion animation produced by manipulating single carbon monoxide molecules.

These breakthroughs piqued the interest of a new generation of scientists.

Professor Feynman's vision was beginning to come to life. Another of his dreams was to construct a quantum computer capable of simulating the quantum universe, which he suggested nearly 40 years ago at an MIT lecture in 1981.

These are only two examples of pivotal moments in the history of quantum thought.

The physicists involved were visionary souls who forced us into new ways of thinking that have eventually culminated in scientific quantum advancements beyond their creators' imaginations.

Chapter 3: MYTHS ABOUT THE QUANTUM UNIVERSE

The laws of physics seemed to be perfectly deterministic for decades. You could predict where every particle would be and what they'd be doing in the future if you knew where they were, how fast they were going, and what forces were between them at any given time. The laws that governed the Universe, from Newton to Maxwell, had no built-in, inherent ambiguity in any sort. Your only limitations stemmed from your lack of experience, measurements, and computing ability.

Much of this changed a little more than a century ago. From radioactivity to the photoelectric effect to the conduct of light as it passes through a double slit, we realized that we could only foresee the likelihood of different outcomes as a result of our Universe's quantum existence in many situations. However, several myths and misconceptions have emerged as a result of this modern, counterintuitive view of truth. The real science behind ten of them is revealed here.

By constructing a track with the outside magnetic rails pointing in one direction and the inside magnetic rails pointing in the opposite direction, you may

1.) Quantum effects can only be observed at very small scales. We generally think about quantum effects in terms of individual particles (or waves) and the peculiar properties they exhibit. However, large-scale, macroscopic effects do occur that are essentially quantum.

When conducting metals are cooled below a certain temperature, they become superconductors, with zero resistance. Building superconducting tracks, in which magnets levitate above and move through them without ever slowing down, is a popular student science project these days, based on quantum effects.

On big, macroscopic scales, superfluids can be formed, as can quantum drums that vibrate both ways. Six Nobel Prizes have been awarded for various macroscopic quantum phenomena over the last 25 years.

2.) "Quantum" is synonymous with "discrete." The idea of chopping up matter (or energy) into individual chunks — or quanta — is important in physics, but it doesn't completely capture what it means to be "quantum" in nature. Take, for example, an atom. Atoms are made up of atomic nuclei that are surrounded by electrons.

Consider the following question: where is the electron at any given time?

Although the electron is a quantum body, its location is unknown until it is measured. When you tie a large number of atoms together (as in a conductor), you'll often find that, although the electrons occupy distinct energy levels, their locations can be anywhere inside the conductor. Many quantum effects are constant, and it's conceivable that space and time, at their most fundamental level, are as well.

By separating two entangled photons from a pre-existing system over a wide distance,...

3.) Information can fly faster than light thanks to quantum entanglement. Here's an experiment that we can do:
- make two intertwined particles
- they are separated by a large distance,
- on your end, measure certain quantum properties (such as the spin) of a single particle
- and you will instantly learn more about the quantum state of another particle: faster than the speed of light.

But here's the thing: no information is transmitted faster than the speed of light in this experiment. Simply put, by calculating the state of one particle, you are constraining the likely outcomes of the other particle. If anyone goes to measure the other particle, they would have no way of knowing that the entanglement has been broken since the first particle has been measured. The only way to know whether entanglement has been broken or not is to reassemble the effects of both measurements: a phase that can only happen at light speed or slower. A 1993 theorem proved that no knowledge can fly faster than light.

4.) Quantum physics is based on the principle of superposition. Consider the possibility of a device being in several quantum states. It could be in state "A" with a 55 percent chance, state "B" with a 30 percent chance, and state "C" with a 15 percent chance. When you make a calculation, though, you'll never see a combination of these potential states; instead, you'll only get a single-state result: "A," "B," or "C."

Superpositions are extremely useful as intermediate calculational measures in evaluating what the potential outcomes (and their probabilities) are, but they are difficult to calculate directly. Furthermore, superpositions do not apply uniformly to all measurables; for example, a superposition of momenta but not positions, or vice versa, is possible. Superposition is not quantifiable or uniformly observable, unlike entanglement, which is a fundamental quantum phenomenon.

Interpretation	Author(s)	Deterministic?	Wavefunction real?	Unique history?	Hidden variables?	Collapsing wavefunctions?	Observer role?	Local?	Counterfactual definiteness?	Universal wavefunction exists?
Ensemble interpretation	Max Born, 1926	Agnostic	No	Yes	Agnostic	No	No	No	No	No
Copenhagen interpretation	Niels Bohr, Werner Heisenberg, 1927	No	No	Yes	No	Yes	Causal	No	No	No
de Broglie–Bohm theory	Louis de Broglie, 1927; David Bohm, 1952	Yes	Yes	Yes	Yes	Phenomenological	No	No	Yes	Yes
Quantum logic	Garrett Birkhoff, 1936	Agnostic	Agnostic	Yes	No	No	Interpretational	Agnostic	No	No
Time-symmetric theories	Satosi Watanabe, 1955	Yes	Yes	Yes	Yes	No	No	No	Yes	Yes
Many-worlds interpretation	Hugh Everett, 1957	Yes	Yes	No	No	No	No	Yes	Ill-posed	Yes
Consciousness causes collapse	Eugene Wigner, 1961	No	Yes	Yes	No	Yes	Causal	No	No	Yes
Stochastic interpretation	Edward Nelson, 1966	No	No	Yes	Yes	No	No	No	Yes	No
Many-minds interpretation	H. Dieter Zeh, 1970	Yes	Yes	No	No	No	Interpretational	No	Yes	Yes
Consistent histories	Robert B. Griffiths, 1984	No	No	No	No	No	No	Yes	No	No
Transactional interpretation	John G. Cramer, 1986	No	Yes	Yes	No	Yes	No	No	Yes	No
Objective collapse theories	Ghirardi–Rimini–Weber, 1986; Penrose interpretation, 1989	No	Yes	Yes	No	Yes	No	No	No	No
Relational interpretation	Carlo Rovelli, 1994	Agnostic	No	Agnostic	No	Yes	Intrinsic	Yes	No	No
QBism	Christopher Fuchs, Ruediger Schack, 2010	No	No	Agnostic	No	Yes	Intrinsic	Yes	No	No

5.) There's nothing wrong with each of us picking our quantum interpretation. Physics is all about predicting, observing, and measuring what you might find in the Universe. However, in quantum mechanics, there are a variety of ways to think about what's going on at the quantum level that all agree with experiments.

- an infinite ensemble of quantum waves, where a measurement selects one member of the ensemble,
- an infinite ensemble of quantum waves, where a measurement selects one member of the ensemble,
- a "quantum handshake" involving a superposition of forward-moving and backward-moving potentials.
- an infinite number of possible worlds corresponding to the possible outcomes, in which we occupy only one direction

as well as a slew of others Choosing one view over another, on the other hand, teaches us nothing but our human prejudices. It's preferable to learn what we can observe and test under different conditions, which is physically true, rather than choosing an understanding that has no experimental advantage over others.

Quantum teleportation is a phenomenon that has been mistakenly marketed as faster-than-light flight. No, in fact...

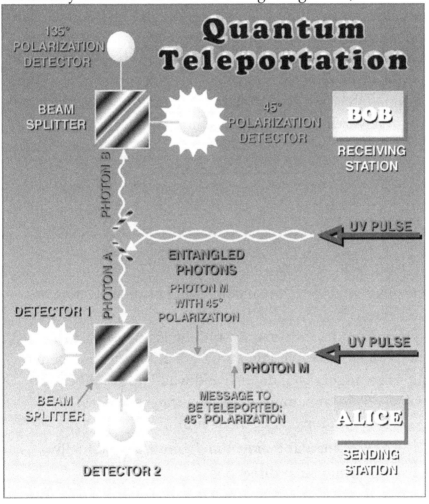

6.) Due to quantum mechanics, teleportation is possible. There is a real concept known as quantum teleportation, but it does not imply that teleporting a physical entity from one place to another is physically possible. You can teleport information from an unknown quantum state on one end to the other end by keeping one entangled particle nearby while sending the other to a desired distinction.

However, there are major drawbacks, such as the fact that it only functions with single particles and that only knowledge about an indeterminate quantum state, not physical matter, can be teleported. Even if you could scale this up to transmit the quantum information that encodes an entire human being, information is not the same as matter: quantum teleportation would never be able to teleport a human.

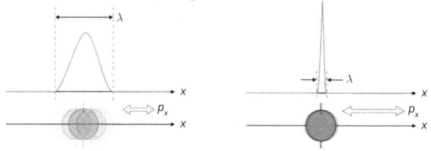

The inherent uncertainty relationship between position and momentum is depicted in this diagram.

7.) In a quantum universe, everything is unknown. In a quantum Universe, certain things are unknown, but many things are highly well-defined and well-known. When you look at an electron, for example, you can't say if it's:

- its position and its momentum,
- or its angular momentum in multiple, mutually perpendicular directions

in any conditions, exactly and simultaneously However, there are certain aspects of the electron that can be precisely known! With absolute certainty, we can determine its rest mass, electric charge, and lifetime (which seems to be infinite). In quantum mechanics, the only unknown factors are pairs of physical quantities that have a particular relationship between them: conjugate variables. This is why energy and time, voltage and free charge, and angular momentum and angular position all have uncertain relationships. Although certain pairs of quantities have inherent uncertainty, many quantities are still known precisely.

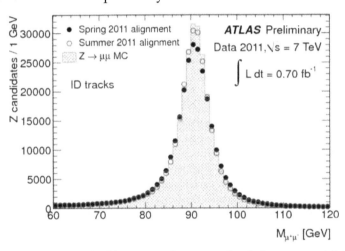

When you're halfway to the top, the inherent width, or half the width of the peak in the above picture,...

8.) The mass of all particles of the same form is the same. If two similar particles, such as two protons or two electrons, were placed on a perfectly accurate scale, they would both have the same mass. Protons and electrons, on the other hand, are stable particles with infinite lifetimes.

Instead, you wouldn't get the same values if you placed unstable particles that decayed after a short time — such as two top quarks or two Higgs bosons — on a perfectly accurate scale. This is due to the inherent ambiguity of energy and time: if a particle only lives for a finite amount of time, the amount of energy (and thus, from E = mc2, rest mass) that the particle has is also unknown. This is referred to as a particle's "width" in particle physics, and it can cause a particle's intrinsic mass to be unknown by up to a few percent.

In the home of Paul Ehrenfest in... [+] Niels Bohr and Albert Einstein debating a wide variety of topics.

9.) Einstein was a staunch opponent of quantum mechanics. It's real that Albert Einstein famously said, "God does not play dice with the Universe." However, arguing against quantum mechanics' inherent randomness — which was the background of that quote — is an argument about how to view quantum mechanics, not an argument against quantum mechanics itself.

In reality, Einstein argued that there could be more to the Universe than we can currently see and that if we could grasp the rules we haven't yet discovered, what appears to us to be randomness might expose a deeper, non-random truth. Although this position has yielded no useful results, research into the basics of quantum physics continues to be a hot topic, with a variety of interpretations involving "hidden variables" in the Universe being successfully ruled out.

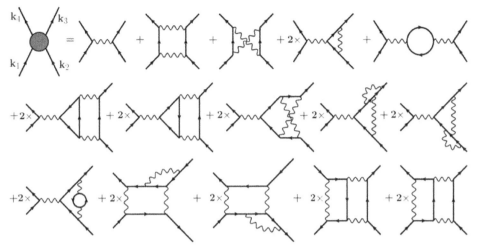

10.) In quantum field theory, particle exchanges fully explain our Universe. The method we most commonly use for measuring the interactions of any two quantum particles is the "dirty little secret" of quantum field theory that physicists study in graduate school. We imagine them as particles being exchanged between those two quanta, as well as any subsequent exchanges that may take place as intermediate steps.

You'd end up with nonsense if you extrapolated this to all imaginable interactions, or what scientists call random loop-orders. This is just a rough approximation: an asymptotic, non-convergent sequence that breaks down after a certain number of terms. It's a very helpful example, but it's essentially incomplete. Digital particle exchanges are an interesting and intuitive concept, but they are unlikely to be the final solution.

THE MYTH OF THE BEGINNING OF TIME

Was the big bang the start of everything? Or was there a time when the world didn't exist? About a decade ago, such a question would have appeared sacrilegious. Most cosmologists said that thinking about a time before the big bang was akin to asking for directions to a location north of the North Pole. However, advancements in theoretical physics, especially the rise of string theory, have shifted their viewpoint. The pre-big bang universe has emerged as cosmology's most recent frontier.

The newest swing of an intellectual pendulum that has swung back and forth for centuries is the ability to explore what could have happened before the big bang. The question of the ultimate beginning has occupied philosophers and theologians in virtually every community in some way or another. It is intertwined with a broad range of issues, one of which is famously encapsulated in a painting by Paul Gauguin from 1897: Do you want to come with us? So, who are we? Do you want to join us? What is our origin story? What exactly are we? What are our plans? The piece portrays the cycle of birth, life, and death, as well as each individual's origin, identity, and destiny, and these personal concerns are linked to celestial ones. We may trace our ancestors back through the centuries, to early forms of life and proto life, to elements synthesized in the primordial universe, and too amorphous energy deposited in space before that. Is there a limit on how far back our family tree can go? Or does its history come to an end? Is the world as finite as we are?

The origin of time was a contentious issue among the ancient Greeks. Taking the no-beginning hand, Aristotle invoked the idea that nothing comes from nothing. It must have always existed if the world could never have gone from nothingness to somethingness. Time must extend eternally into the past and future for this and other purposes. Christian theologians, on the other hand, tended to hold the opposing viewpoint. Augustine said that God resides outside of space and time and that he is capable of creating these structures just as he is capable of creating other facets of our universe. When asked, "What was God doing before he created the world?" he responded, "What was God doing before he created the world?" There was no before, Augustine replied, since time is a part of God's existence.

Modern cosmologists have come to the same conclusion as Albert Einstein's general theory of relativity. Space and time, according to the theory, are soft and malleable entities. Space is inherently dynamic on the largest scales, expanding and contracting over time and bringing matter like driftwood on the tide. In the 1920s, scientists confirmed that our universe is expanding, with distant galaxies moving away from one another. As physicists Stephen W. Hawking and Roger Penrose demonstrated in the 1960s, time cannot be extended forever. When you play cosmic history backward in time, all of the galaxies collide into a single infinitesimal point known as a singularity, just like they're falling into a black hole. Each galaxy, or its predecessor, is shrunk to a minuscule scale. Density, temperature, and the curvature of spacetime become infinite. Our celestial ancestors cannot reach past the singularity, which is the supreme cataclysm.

For cosmologists, the unavoidable singularity faces severe challenges. It clashes with the universe's high degree of homogeneity and isotropy on large scales. Some kind of communication had to move among distant regions of space, organizing their properties, for the universe to look broadly the same everywhere. However, such correspondence runs counter to the old cosmological model.

Consider what has happened in the 13.7 billion years since the cosmic microwave background radiation was first released. Because of the expansion, the distance between galaxies has increased by a factor of about 1,000, while the radius of the observable universe has increased by a factor of about 100,000. (because light outpaces the expansion). We can now see areas of the universe that were not apparent 13.7 billion years ago. Indeed, light from the farthest galaxies has entered the Milky Way for the first time in galactic history.

Strange Coincidence

Regardless, the Milky Way's properties are essentially the same as those of distant galaxies. It's as if you turned up at a party only to discover that you were dressed identically to a dozen of your closest peers. It could be explained away as a coincidence if only two of you were dressed the same, but a dozen indicates that the partygoers had prepared their costumes ahead of time. The number of cosmology is tens of thousands, not a dozen—the number of independently similar patches of sky in the microwave background.

One hypothesis is that all of those regions of space were born with similar properties, meaning that the homogeneity is merely coincidental. Physicists, on the other hand, have proposed two more natural ways out of the impasse: the early universe was much smaller or older than standard cosmology suggests. Intercommunication would have been possible with one (or both, working together).

Following the first option is the most common option. It proposes that early in the universe's history, the universe went through a time of accelerating expansion known as inflation. Galaxies or their precursors were so close together before this process that they could easily coordinate their properties. They lost touch during inflation because light couldn't keep up with the frantic growth. After inflation stopped, the expansion slowed, and galaxies eventually re-entered each other's vision.

The theoretical energy contained in a new quantum field, the inflation, around 1035 seconds after the big bang, is attributed to the inflationary spurt by physicists. Gravitational repulsion is caused by potential energy rather than rest mass or kinetic energy. The inflaton accelerated the expansion rather than slowing it down as ordinary matter's gravitation would. The Inflationary Universe, by Alan H. Guth and Paul J. Steinhardt; SCIENTIFIC AMERICAN, May 1984; and Four Keys to Cosmology, Special report; SCIENTIFIC AMERICAN, February 2004] [see The Inflationary Universe, by Alan H. Guth and Paul J. Steinhardt; SCIENTIFIC AMERICAN, May 1984; and Four Keys to Cosmology, Special report; SCIENTIFIC AMERI However, there are several potential theoretical issues, starting with the question of what the inflaton was and what gave it such a large initial potential energy.

A less well-known method of solving the puzzle is to remove the singularity, as stated in the second choice. If time didn't start with the big bang, and a long period followed the current galactic expansion, the matter would have had plenty of time to organize itself smoothly. As a result, scientists have reexamined their reasons for inferring a singularity.

One of the assumptions is debatable: relativity theory is still true. Quantum effects would have been important, if not dominant, close to the putative singularity. Since standard relativity ignores such results, believing the singularity's inevitability entails trusting the theory beyond reason. Physicists would subsume relativity in a quantum theory of gravity to figure out what happened. From Einstein onwards, the challenge has occupied theorists, but progress was slow until the mid-1980s.

Evolution of a Revolution

TWO APPROACHES STAND OUT TODAY. One, known as loop quantum gravity, keeps Einstein's principle intact but modifies the method for applying it in quantum mechanics [see page 82 of Lee Smolin's Atoms of Space and Time]. Over the last few years, loop quantum gravity practitioners have made significant progress and gained profound insights. Nonetheless, their solution may not be radical enough to overcome the fundamental problems of quantizing gravity. After Enrico Fermi introduced his powerful theory of the weak nuclear force in 1934, particle theorists faced a similar challenge. Attempts to construct a quantum version of Fermi's theory have all failed miserably. The deep modifications brought about by Sheldon L. Glashow, Steven Weinberg, and Abdus Salam's electroweak theory in the late 1960s were what was needed, not a new technique.

String theory, a genuinely groundbreaking modification of Einstein's theory, is the second solution, which I believe is more promising. While proponents of loop quantum gravity claim to draw several of the same conclusions, this BOOK will concentrate on it.

String theory evolved from a model I devised in 1968 to describe the world of nuclear particles (like protons and neutrons) and their interactions. The model failed, despite the initial excitement. It was replaced by quantum chromodynamics, which describes nuclear particles in terms of their more elementary constituents, quarks, several years later. Quarks are held together by elastic strings inside a proton or neutron. The original string theory, in retrospect, had captured those stringy aspects of the nuclear world. It was only later that it was resurrected as a contender for combining general relativity and quantum theory.

The basic idea is that elementary particles are infinitely thin one-dimensional objects called strings, rather than pointlike objects. The large zoo of elementary particles, each with its own set of properties, reflects the wide range of possible string vibration patterns. How can a simplistic theory adequately describe the complex world of particles and their interactions? Quantum string magic, as it's known, holds the key to the solution. When the laws of quantum mechanics are applied to a vibrating string—which looks like a miniature violin string but vibrates at the speed of light—new properties emerge. All of these discoveries have far-reaching implications for particle physics and cosmology.

Quantum strings, for starters, have a finite size. A violin string could be cut in half, cut in half again, and so on until it became a massless pointlike particle if it weren't for quantum effects. However, the Heisenberg uncertainty principle finally kicks in, preventing the lightest strings from being cut any less than 1034 meters. String theory introduced this irreducible quantum of length, denoted ls, as a new natural constant alongside the speed of light, c, and Planck's constant, h. It's important in almost every aspect of string theory since it sets a finite limit on quantities that might otherwise go to zero or infinite.

Second, even though quantum strings are massless, they can have angular momentum. Angular momentum is a property of an object that rotates around an axis in classical physics. Since angular momentum is calculated by multiplying velocity, mass, and distance from the axis, a massless object cannot have angular momentum. Quantum variations, on the other hand, alter the condition. Without acquiring any mass, a tiny string can gain up to two units of h of angular momentum. This function is extremely useful because it perfectly matches the properties of all known fundamental forces' carriers, such as the photon (for electromagnetism) and the graviton (for gravity) (for gravity). Historically, it was angular momentum that first alerted physicists to string theory's quantum-gravitational consequences.

Third, quantum strings necessitate the presence of additional spatial dimensions outside the standard three. A quantum string is more finicky than a classical violin string, which can vibrate regardless of the properties of space and time. Unless spacetime is strongly curved (contrary to observations) or comprises six extra spatial dimensions, the equations explaining the vibration become incongruent.

Fourth, physical constants no longer have arbitrary, defined values, such as Newton's and Coulomb's constants, which occur in physics equations and define the properties of nature. They appear in string theory as fields that can change their values dynamically, similar to the electromagnetic field. These fields may have had different values in different cosmological epochs or far-flung parts of space, and the physical constants may still vary slightly today. Observing even the tiniest variance would give string theory a major boost.

The master key to string theory is a field called the dilaton, which measures the total strength of all interactions. String theorists are intrigued by the dilaton because its value can be reinterpreted as the scale of an extra space dimension, resulting in a total of 11 spacetime dimensions.

Tightening the Loose Ends

FINALLY, QUANTUM strings have introduced physicists to some striking new natural symmetries known as dualities, which change our understanding of what happens when things get very small. I've already stated a duality: a short string is usually lighter than a long one, but if we try to shrink it below the fundamental length ls, the string becomes heavier again.

T-duality, another form of symmetry, states that small and large extra dimensions are equivalent. Strings can travel in more complicated ways than pointlike objects, resulting in this symmetry. Consider a closed string (a loop) on a cylindrically shaped space with one finite extra dimension represented by its circular cross-section. Aside from vibrating, the string may wrap around the cylinder in a single or multiple loops, similar to a rubber band wrapped around a rolled-up poster.

The energetic cost of these two string states is proportional to the cylinder's size. Winding energy is proportional to the radius of the cylinder: larger cylinders cause the string to stretch more as it wraps around, so the windings contain more energy than on a smaller cylinder. Moving around a circle, on the other hand, is inversely proportional to its radius: larger cylinders allow for longer wavelengths (smaller frequencies), which reflect less energy than shorter wavelengths. The two states of motion will switch roles if a large cylinder is substituted for a small one. Winding produces energies that were previously provided by circular motion and vice versa. An observer from the outside only notices the energy levels, not the source of those levels. The large and small radii are physically similar to the observer.

Although T-duality is most commonly defined in terms of cylindrical spaces in which one dimension (the circumference) is finite, a version of it may also be applied to our ordinary three dimensions, which tend to extend indefinitely. When discussing the infinite expansion of space, one must exercise caution. Its overall size is unchangeable; it is infinite. It will, however, extend in the sense that bodies embedded within it, such as galaxies, move apart. The most important variable is the scale factor, which is the factor by which the distance between galaxies varies, manifesting itself as the galactic redshift observed by astronomers. Universes with small-scale factors are similar to universes with large-scale factors, according to T-duality. Einstein's equations lack such symmetry; it comes from the unification that string theory embodies, with the dilaton playing a key role.

For a long time, string theorists believed that T-duality only applied to closed strings, not open strings, which have loose ends and thus cannot wind. T-duality did apply to open strings, according to Joseph Polchinski of the University of California, Santa Barbara, given that the transition between large and small radii was followed by a shift in the conditions at the string's endpoints. Physicists had previously proposed boundary conditions in which no force was applied to the ends of the strings, allowing them to flap around freely. These conditions become so-called Dirichlet boundary conditions under T-duality, in which the ends remain put.

Both types of boundary conditions can be mixed in any series. Electrons, for example, maybe strings of ends that can travel freely in three of the ten spatial dimensions but are trapped in the other seven. A Dirichlet membrane, or D-brane, is a subspace formed by those three dimensions. Petr Horava of the University of California, Berkeley, and Edward Witten of Princeton's Institute for Advanced Study suggested in 1996 that our universe exists on such a brane. We are unable to perceive the absolute 10-dimensional majesty of space due to the partial mobility of electrons and other particles.

All of quantum strings' magical properties point in one direction: they despise infinity. They can't fall to an infinitesimal point, so they don't have to deal with the paradoxes that come with it. Their nonzero size and novel symmetries impose upper bounds on physical quantities that would otherwise grow indefinitely in traditional theories and lower bounds on quantities that decrease. If one replays the history of the universe backward in time, string theorists predict that the curvature of spacetime will increase. Instead of expanding to infinity (as in the conventional big bang singularity), it ultimately reaches a limit and contracts again. Before string theory, physicists struggled to conceive a process that could eradicate the singularity so completely.

Taming the Infinite

The conditions at the big bang's zero point were so severe that no one has yet found out how to solve the equations. Despite this, string theorists have made educated guesses about the universe before the big bang. Two popular models are floating around.

The first, known as the big bang scenario, which my colleagues and I started working on in 1991, blends T-duality with the more well-known symmetry of time reversal, in which physics equations function equally well when applied backward and forward in time. The combination leads to new possible cosmologies in which the universe expanded at the same rate five seconds before the big bang as it did five seconds after the bang. However, the rate of change of the expansion was the polar opposite at the two points: if it was slowing after the bang, it was speeding up before. In other words, the big bang may have been a violent transition from acceleration to deceleration rather than the beginning of the universe.

The beauty of this example is that it immediately integrates a crucial insight from standard inflationary theory: the universe had to accelerate to become so homogeneous and isotropic. Acceleration occurs after the big bang, according to standard theory, due to an ad hoc inflaton field. In the big bang case, it happens before the big bang as a natural result of string theory's novel symmetries.

The pre-bang universe was almost a complete mirror image of the post-bang universe, according to the scenario [see box on page 77]. If the world is infinite in the future, with its contents dwindling to a gruel, it is also eternal in the past. It was nearly empty Infinitum ago, containing only a tenuous, widely scattered, chaotic gas of radiation and matter. The dilaton field dominated nature's forces so weakly that particles in this gas scarcely interacted.

The forces grew stronger over time and began to pull matter together. Some areas accumulated matter at the expense of their surroundings at random. The density in these areas eventually became so high that black holes began to form. The matter inside those regions was then cut off from the rest of the universe, shattering it into fragments.

Space and time are inverted inside a black hole. The black hole's nucleus is not a point in space, but a point in time. The infalling matter achieved higher and higher densities as it entered the heart. However, when the density, temperature, and curvature reached the maximum values permitted by string theory, these quantities bounced back and began to fall. The point at which the reversal occurs is referred to as a big bang. Our universe emerged from the interior of one of those black holes.

Unsurprisingly, such an out-of-the-box situation has sparked debate. According to Andrei Linde of Stanford University, the black hole that gave rise to our universe would have to have developed with an extraordinarily large size—much greater than the length scale of string theory—for this scenario to fit observations. The equations predict black holes of all sizes, which is a response to this objection. Our universe evolved inside a sufficiently large one.

A more severe objection, advanced by Thibault Damour of the Institut des Hautes tudes Scientifiques in Bures-sur-Yvette, France, and Marc Henneaux of the Free University of Brussels, is that matter and spacetime may have behaved chaotically near the big bang, possibly contradicting the early universe's observed regularity. A chaotic state, I recently suggested, would create a dense gas of miniature string holes—strings so small and huge that they were on the verge of being black holes. Damour and Henneaux identified a problem, which could be solved by the behavior of these holes. Thomas Banks of Rutgers University and Willy Fischler of the University of Texas at Austin also suggested a similar concept. Other objections exist, although it remains to be seen if they have found a fatal flaw in the scenario.

The ekpyrotic (conflagration) scenario is another popular model for the universe before the big bang. The ekpyrotic scenario, which was developed five years ago by a group of cosmologists and string theorists including Justin Khoury of Columbia University, Paul J. Steinhardt of Princeton University, Burt A. Ovrut of the University of Pennsylvania, Nathan Seiberg of the Institute for Advanced Study, and Neil Turok of the University of Cambridge, is based on the previously mentioned Horava-Witten idea that our universe sits on the edge of the known universe. The two branes attract each other and collide now and then, causing the extra dimension to shrink to zero before growing again. The time of collision would correspond to the big bang.

Collisions occur cyclically in a variant of this scenario. Two branes may collide, bounce off one another, move apart, pull together, collide again, and so on. The branes behave like Silly Putty in between collisions, expanding as they recede and contracting slightly as they reassemble. The rate of expansion accelerates during the turnaround; indeed, the universe's current accelerated expansion may portend another collision. There are some similarities between the big bang and ekpyrotic scenarios. Both begin with a vast, cold, nearly empty universe, and both face the difficult (and unsolved) task of transitioning from the pre- to post-bang phase. The behavior of the dilaton field is the main mathematical difference between the scenarios. The dilaton begins with a low value in the big bang, indicating that the forces of nature are weak, and gradually increase in strength. The ekpyrotic scenario, in which the collision occurs when the forces are at their weakest, is the polar opposite.

The ekpyrotic theory's creators hoped that the forces' weakness would make it easier to analyze the bounce, but they were still faced with a difficult high-curvature situation, so the jury is still out on whether the scenario truly avoids a singularity. Also, to solve the usual cosmological puzzles, the ekpyrotic scenario must include very specific conditions. For example, the branes about to collide had to be nearly parallel to one another, or the collision would not have resulted in a sufficiently homogeneous bang. Because successive collisions allow the branes to straighten themselves, the cyclic version may be able to solve this problem.

Chapter 4: UNDERSTANDING FREQUENCIES

You're onstage or in the studio, and you're trying to describe a sound to the engineer in charge of whatever project you're working on. It's at this point that you're at a loss for words as to how to describe the synth's high-pitched, squeaky sound, or the guitar's flabby, meedley-deedley sound. How do you put something into words that defies description, and perhaps even imitation? This quick guide will introduce you to the world of sounds, as well as how to refer to and discuss them. What is a frequency?

When we talk about sounds in the audio world and where they live in pitch (low to high), we usually refer to them in terms of frequency. Sound is a wave, which is a movement of air molecules that our brain converts to sound via a surprisingly complicated series of internal workings in our ears. The number of times these waves complete a cycle in a second can be used to calculate their size. (Does that day in high school physics class seem to be resurfacing now?) These cycles per second are measured in hertz, which is a unit of measurement (Hz). The reference pitch A440, which is 440 Hz, is used in music, particularly in tuning. This is the note that produces a vibration at 440 cycles per second.

So, now that we've figured out what these numbers and notes mean, what's next?

Human hearing range

Human hearing ranges from 20 Hz to 20,000 Hz, according to widely accepted standards (or 20k Hz). While most of us are born with this frequency range, most adults have a 20 Hz to 15k or 16k Hz frequency range (barring no high-frequency-specific hearing loss). Even though this sounds astronomical, the frequency scale is not evenly divided. To go up an octave, for example, double the frequency; to go down an octave, halve it. On the piano, the A above middle C is 440 Hz, and the A an octave above it is 880 Hz, but the A below middle C is 220 Hz. This means that between 10,000 and 20,000 Hz, there is only one octave (12 half-steps) of notes, and between 80 and 160 Hz, there is only one octave.

We now have a better understanding of how sounds are measured and what the playing field for what we can hear is. But how can we put these sounds into words?

Hz: 20–80

This is the true bottom end of the spectrum. This range's bottom half (20 Hz to 40 Hz) is more felt than heard. It's difficult to tell if something is true or not in this range. Most speaker systems, including high-end studio monitors, don't even produce sound accurately in this range. For comparison, the note F0 (21.8 Hz fundamental) on an Imperial Bosendorfer extended grand piano and the note A0 (27.5 Hz fundamental) on a standard concert grand are both difficult to tune at their fundamentals. The four-string bass's lowest note (fundamental E at 41 Hz) is played in the upper half (40 Hz to 80 Hz). When you hear it, you'll get a rumbly sensation in your chest.

80 to 160 Hz

This is the point at which we enter what is commonly referred to as the bass range. Most consumer-grade mixers with fixed EQ points and home stereos set their "low" band between 80 and 120 Hz. The guitar now enters the spectrum (the fundamental pitch of the low E string in standard tuning is 82.5 Hz), and the bass begins its exit at its fundamental pitch (G string open fundamental is 98 Hz). When this range is increased, things can feel boomy or thumpy, but it also adds warmth. The big kick you get in a dance club when the beat is thumping away, for example, is usually around 100 to 120 Hz. Low-end instruments (bass, kick drum, piano, synths) that don't have enough of this range can make them feel thin and anemic. This is where you'll find powerful, rumbling, low-sounding feedback from monitors on stage.

People who aren't used to thinking about sound in terms of frequencies mistakenly believe that low frequencies are lower and high frequencies are higher than they are.

Hz range: 160 to 500

This range covers a lot of ground. Many people who aren't used to thinking about sound in terms of frequencies mistakenly believe that low frequencies are lower and high frequencies are higher. The guitar begins to fade away at its fundamental frequency in this range (high E string open fundamental is 330 Hz). However, the frequency range of 200 to 250 Hz is a double-edged sword; it is here that things sound really warm and sweet, but too much of it gives you that muddy feeling, similar to when you have a cold and your voice sounds muffled in your head. Simply put, a 200 Hz build-up is a head cold. Above this, between 250 and 500 Hz, things can start to sound boxy (yes, this is a commonly accepted term). Imagine hitting or knocking on a hollow box and hearing a woody ring. It's not quite as muddy and low as the "head cold," but it's close. Here's where you'll look for problems with that.

500 to 1.6k Hz

In terms of consumer EQ, we're now in the mid-to-upper-mids range. The guitar's fundamental pitch is completely out, with the highest frets around 900 Hz. Too much in the 500-900 Hz range can make things honky or nasal. The teacher from the Charlie Brown cartoons serves as the audio aid for this range. The feeling you get when you hear "Wha Wha" is how a build-up will feel in this band. Above this, you'll hear sounds that become more pointed and sharp-sounding, such as the sound that goes with the TV test pattern (imagine your local public access channel when it's turned off). The sound that goes along with it, the "beeeeep," is a 1,000 Hz pure sine wave. So if you hear something that irritates you the same way, or feedback that sounds similar in pitch, you're now in the 1k Hz to 1.6k Hz range.

1.6 to 4k Hz

The "presence" of the human voice is found in this range. A boost in this range (usually around 3k Hz) will liven things up if they sound dull or flat. However, we encounter the Goldilocks scenario, which all sound engineers face: too much boost in this range causes sounds to become harsh and edgy. We also lose the piano's fundamental pitches at this point, with the highest keys typically peaking at around 4k Hz. Guitars share the same presence element as the human voice, and they frequently compete for the same sonic territory. There's a reason lead vocalists and lead guitarists are often at odds, not only in terms of stage presence but also in terms of sonic space.

4 to 10k Hz

Welcome to the upper echelon of society. Somewhere up here, our home hi-fi and consumer high EQs are stored. (Keep in mind that we're only dealing with about an octave here.) Up here, the sounds are more of the hiss and squeal variety – you know, the ones that hurt. This range is dominated by sibilants, such as the S's in words. Things sound undefined or lack a certain crispness without them. Around 7k Hz to 10k Hz, cymbals and other percussion produce a sizzle. This band can deal with shrieking, piercing feedback, or a real crunchy, tinny quality to sounds.

10k Hz and beyond

These are the apex of the peaks. This is when frequency response begins to drop out, just as it did at the low end, but for the opposite reasons. Sometimes it's because a microphone's transducer can't accurately respond to these frequencies, and other times it's simply because people can't hear what's going on in this range. (In this field, high-end hearing is usually the first to go.) The best way to describe these frequencies is "air." The presence of overtones in this range is usually responsible for a sound's heady, open quality. Before you go pegging out all of your EQs at 10 to 12k Hz to add airiness, keep in mind that simply boosting this range will result in noise if nothing exists there, to begin with. Jingle a set of car keys, for example. In this air range, we refer to the crisp, bell-like quality of the keys hitting one another. Can't seem to hear it? Don't worry, I know a few fantastic mix engineers who are nearly deaf to anything above 14k Hz and still produce fantastic work.

I hope that by breaking down what you hear and how you hear it down, you will be able to better describe what you hear and how you hear it. At the very least, you now know why some sound engineers roll their eyes when you request more "highs" in your monitor mix – it's not that simple.

Chapter 5: THREE WAYS QUANTUM PHYSICS AFFECTS YOUR DAILY LIFE

Quantum physics is arguably the greatest intellectual achievement in human history, but it appears to most people to be too far away and abstract to matter. This is largely a self-inflicted wound on the part of physicists and popular science writers: when we talk about quantum physics, we usually emphasize the strange and counter-intuitive phenomena: Schrödinger's cat in a state of "alive" and "dead," Einstein's objection to God playing dice, and quantum entanglement's strange long-distance correlations. These phenomena are fascinating because they are unusual, but studying them in the lab necessitates isolating very simple quantum systems, and it can be difficult to see any link between them and everyday life.

Quantum physics, on the other hand, is all around us. The universe as we know it is governed by quantum rules, and while the classical physics that emerges when quantum physics is applied to enormously large numbers of particles appears to be very different, quantum effects are responsible for many familiar, everyday phenomena. Here are a few examples of quantum phenomena you're likely to encounter in your daily life without realizing it:

close-up of a young man drinking coffee and toasting his bread from the toaster

Toasters: Most of us are familiar with the red glow of a heating element as we toast a slice of bread or a bagel. It was also the birthplace of quantum mechanics: Quantum physics was created to solve the problem of explaining why hot objects glow a specific color of red.

The color of light emitted by a hot object is an example of the kind of simple, universal phenomenon that theoretical physicists love: no matter what an object is made of, if it can withstand being heated to a certain temperature, the spectrum of light it emits is identical to that of any other substance. In the late 1800s, this kind of universal behavior drew a lot of very bright physicists, but none of them were able to solve the problem.

The fact that light was unaffected by composition suggested a simple universal approach: add up all the colors of light that an object might emit and give each one an equal share of the object's heat energy.

The problem is that there are many more ways to emit high-frequency light than low-frequency light, implying that your toaster should be spraying x-rays and gamma rays all over the kitchen instead of a pleasant warm res glow. Something else must be going on because that isn't happening (which is a good thing!).

Max Planck discovered the solution to this problem when he proposed the "quantum hypothesis," which stated that light could only be emitted in discrete chunks of energy, integer multiples of a small constant times the frequency of the light. This energy quantum is larger than the share of heat energy allotted to that frequency for high-frequency light, so no light is emitted at that frequency. This blocks the high-frequency light, resulting in a formula that closely matches the spectrum of light emitted by hot objects.

As a result, every time you toast bread, you're looking at the origins of quantum physics.

One hanging eco-energy saving light bulb glows and stands out from unlit incandescent bulbs. On... [+] dark blue background, leadership, and different creative idea concept, Rendering in 3D.

Fluorescent Lights: Incandescent light bulbs produce light by heating a piece of wire to the point where it emits a bright white glow, making them quantum in the same way that a toaster does. You're getting light from another revolutionary quantum process if you have fluorescent bulbs around—either the long tubes or the newer twisty CFL bulbs.

Physicists discovered that each element in the periodic table has its spectrum in the early 1800s: when atoms are heated, they emit light at a small number of discrete frequencies, with a different pattern for each element. These "spectral lines" were quickly used to determine the composition of unknown materials and even to discover the presence of previously unknown elements—helium, for example, was discovered as a previously unknown spectral line in sunlight.

While this was undeniably effective, no one could explain it until Niels Bohr introduced the first quantum model of an atom in 1913, based on Planck's quantum idea (which Einstein extended in 1905) and Planck's quantum idea (which Einstein extended in 1905) Bohr proposed that an electron can happily orbit the nucleus of an atom in certain special states, and that atoms absorb and emit light only as they move between those states. In the way introduced by Planck, the frequency of light absorbed or emitted is determined by the energy difference between states, resulting in a set of discrete frequencies for any given atom.

This was a revolutionary concept, but it worked brilliantly to explain the spectrum of light emitted by hydrogen, as well as the x-rays emitted by a wide range of elements, and quantum physics was off and running. While today's view of what happens inside an atom differs significantly from Bohr's original model, the basic concept remains the same: electrons move between special states inside atoms by absorbing and emitting light of specific frequencies.

Fluorescent lighting is based on the following principle: A small amount of mercury vapor is excited into a plasma inside a fluorescent bulb (either long tube or CFL). Mercury emits light at frequencies that are most visible in the visible spectrum, fooling our eyes into thinking the light is white. When you look at a fluorescent bulb through a novelty diffraction grating, you'll see a few distinct colored images of the bulb, whereas an incandescent bulb produces a continuous rainbow smear.

As a result, whenever you use fluorescent lights to light your home or office, you can thank quantum physics.

An unidentified young businesswoman works on her laptop in her home office in a high-angle shot.

Computers: While Bohr's quantum model was undeniably useful, it didn't come with a physical explanation for why electrons in atoms should have special states. That didn't happen for almost a decade, but once it did, it became the foundation for the last century's most transformative technological revolution.

Louis de Broglie, a French Ph.D. student from an aristocratic family, came up with the radical idea that provided a physical basis for Bohr's special energy states. He proposed that, just as Planck and Einstein had introduced a particle-like nature for light waves (where a beam of light can be thought of as a stream of "light quanta" each carrying one unit of energy for that frequency), particles like electrons might have a corresponding wave-like behavior. When you give electrons a wavelength that depends on their momentum, you can find "standing wave" orbits in which the electron wave completes an integer number of oscillations as it travels around the nucleus, and these have the exact energies to be Bohr's special states in hydrogen.

This wave behavior can be directly measured, and it was done quickly in both the United States and the United Kingdom. Erwin Schrödinger developed his wave equation as a result of thinking about those waves, and thus one of the main approaches to the full modern theory of quantum mechanics. Our understanding of how electrons move through materials has been profoundly altered by their wave nature, leading to our current understanding of energy bands and band gaps within materials. We can use this physics to control the electrical properties of semiconductors, and we can make tiny transistors that form the basic bits used to process digital information by sticking together bits of silicon with the exact right admixture of other elements.

So, every time you turn on your computer (say, to read a blog post about quantum physics), you're taking advantage of electrons' wave nature and the unprecedented control over materials that this allows. It may not be the most glamorous type of quantum computer, but quantum physics is required for any modern computer to function properly.

These are just a few examples of how quantum physics manifests itself in your daily life. If you'd like to learn more about these topics, as well as a few others that I may discuss in future posts, check out my new book, Breakfast with Einstein, which is available now in the UK from Oneworld Publications and next week in the US from BenBella Books. Just in time for all of your holiday gift-giving needs around the Winter Solstice...

Chapter 6: BLACK HOLES

Some of the strangest and most fascinating objects in space are black holes. They're extremely dense, and their gravitational attraction is so strong that even light cannot escape their grasp if it gets close enough.

With his general theory of relativity, Albert Einstein predicted the existence of black holes in 1916. Many years later, in 1967, American astronomer John Wheeler coined the term "black hole." The first physical black hole was discovered in 1971, after decades of only knowing about them as theoretical objects.

The Event Horizon Telescope (EHT) collaboration then released the first image of a black hole in 2019. While looking at the event horizon, or the area beyond which nothing can escape from a black hole, the EHT discovered the black hole at the center of galaxy M87. The image depicts the loss of photons in a sudden manner (particles of light). It also opens up a whole new field of research into black holes now that astronomers have a better understanding of what they look like.

Astronomers have discovered three different types of black holes so far: stellar black holes, supermassive black holes, and intermediate black holes.

Small but deadly stellar black holes

When a star reaches the end of its fuel supply, it may collapse or fall into itself. The new core of smaller stars (up to three times the mass of the sun) will become a neutron star or a white dwarf. When a larger star collapses, however, the star continues to compress, forming a stellar black hole.

Individual stars collapse into black holes, which are relatively small but extremely dense. One of these objects has more than three times the mass of the sun packed into a city's diameter. This causes a tremendous amount of gravitational force to be exerted on objects in the vicinity of the object. The dust and gas from their surrounding galaxies are then consumed by stellar black holes, causing them to expand in size.

"The Milky Way contains a few hundred million" stellar black holes, according to the Harvard-Smithsonian Center for Astrophysics.

Supermassive black holes — the birth of giants

Small black holes abound in the universe, but supermassive black holes, their cousins, rule. These massive black holes are millions or billions of times as massive as the sun but have a diameter of about the same size. Black holes are thought to exist at the heart of almost every galaxy, including our own Milky Way.

Scientists are baffled as to how such massive black holes form. Once formed, these giants gather mass from the dust and gas that surrounds them, which is abundant in the center of galaxies, allowing them to grow even larger.

Hundreds of thousands of tiny black holes may merge to form supermassive black holes. Large gas clouds, collapsing together and rapidly accumulating mass, could also be to blame. A third possibility is the collapse of a stellar cluster, which is a collection of stars that all collide at the same time. Fourth, large clusters of dark matter could produce supermassive black holes. We can see dark matter because of its gravitational effect on other objects; however, we don't know what it's made of because it doesn't emit light and can't be observed directly.

A young black hole, such as the two distant dust-free quasars discovered recently by the Spitzer Space Telescope, is depicted.

A young black hole, such as the two distant dust-free quasars discovered recently by the Spitzer Space Telescope, is depicted. More photos of black holes of the universe (Image credit: NASA/JPL-Caltech)

Black holes that are stuck in the middle are known as intermediate black holes.

Scientists used to believe that black holes only came in two sizes: small and large, but new research suggests that midsize, or intermediate, black holes (IMBHs) may exist. When stars in a cluster collide in a chain reaction, such bodies could form. Several of these IMBHs forming in the same area could eventually collide in the galaxy's center, forming a supermassive black hole.

In the arm of a spiral galaxy, astronomers discovered what appeared to be an intermediate-mass black hole in 2014.

In a statement, study co-author Tim Roberts of the University of Durham in the United Kingdom said, "Astronomers have been looking very hard for these medium-sized black holes." "There have been hints that they exist, but IMBHs have acted like a long-lost relative who doesn't want to be found."

These IMBHs may exist in the heart of dwarf galaxies, according to newer research from 2018. (or very small galaxies). Observations of ten such galaxies (five of which had previously been unknown to science before this latest survey) revealed X-ray activity, which is common in black holes, indicating the presence of black holes with masses ranging from 36,000 to 316,000 solar masses. The data came from the Sloan Digital Sky Survey, which looks at over a million galaxies and can detect the type of light that is often seen coming from black holes picking up nearby debris.

What do black holes look like?

Strange regions where gravity is strong enough to bend light, warp space, and distort time are known as black holes.

Strange regions where gravity is strong enough to bend light, warp space, and distort time are known as black holes. This SPACE.com infographic explains how black holes work.

The outer and inner event horizons, as well as the singularity, are the three "layers" of a black hole.

The event horizon of a black hole is the boundary that surrounds the black hole's mouth and beyond which no light can escape. Once a particle crosses the event horizon, it cannot leave. Across the event horizon, gravity remains constant.

The singularity, or single point in space-time where the black hole's mass is concentrated, is the inner region of a black hole, where the object's mass is concentrated.

Scientists are unable to see black holes in the same way that they can see stars and other celestial objects. Astronomers must instead rely on detecting the radiation that black holes emit as dust and gas are drawn into the dense creatures.

However, supermassive black holes in the center of galaxies may be obscured by the thick dust and gas that surrounds them, obscuring the telltale emissions.

When matter is drawn toward a black hole, it can ricochet off the event horizon and be hurled outward instead of being drawn into the maw. The material is accelerated to near-relativistic speeds, resulting in bright jets of material. Although the black hole is invisible, these powerful jets can be seen from great distances.

The image of a black hole taken by the Event Horizon Telescope in M87 (released in 2019) took an extraordinary amount of effort, requiring two years of research even after the images were taken. This is because the collaboration of telescopes, which spans many observatories around the world, generates an enormous amount of data that is too large to send over the internet.

Researchers hope to image other black holes in the future and compile a database of what they look like. Sagittarius A*, the black hole at the center of our own Milky Way galaxy, is most likely the next target. According to a 2019 study, Sagittarius A* is intriguing because it is quieter than expected, which could be due to magnetic fields suffocating its activity. Another study that year discovered that Sagittarius A* is surrounded by a cool gas halo, providing unprecedented insight into the environment around a black hole.

This image of the supermassive black hole in the center of the galaxy M87 and its shadow was captured by the Event Horizon Telescope, a planet-scale array of eight ground-based radio telescopes forged through international collaboration.

This image of the supermassive black hole in the center of the galaxy M87 and its shadow was captured by the Event Horizon Telescope, a planet-scale array of eight ground-based radio telescopes forged through international collaboration. (Photo courtesy of EHT Collaboration)

Binary black holes are being illuminated.

Using the Laser Interferometer Gravitational-Wave Observatory (LIGO), astronomers discovered gravitational waves from merging stellar black holes in 2015.

In a statement, David Shoemaker, a spokesperson for the LIGO Scientific Collaboration (LSC), said, "We now have further confirmation of the existence of stellar-mass black holes larger than 20 solar masses — these are objects we didn't know existed before LIGO detected them." The observations made by LIGO also reveal the direction in which a black hole spins. Two black holes can spin in the same direction or the opposite direction as they spiral around one another.

Binary black holes are thought to form in two ways. The first hypothesis proposes that the two black holes formed as a binary at around the same time, from two stars that were born together and died explosively at around the same time. The companion stars would have had the same spin orientation as each other, and the two black holes left behind would have had the same spin orientation as well.

Black holes in a stellar cluster sink to the cluster's center and pair up in the second model. In comparison to one another, these companions would have random spin orientations. The discovery of companion black holes with different spin orientations by LIGO adds to the evidence for this theory.

"We're starting to collect real statistics on binary black hole systems," said Caltech's Keita Kawabe, who works at the LIGO Hanford Observatory. "That's interesting because some models of black hole binary formation are slightly more favored than others even now, and we can narrow this down in the future."

Weird facts about black holes

Gravity would stretch you out like spaghetti if you fell into a black hole, according to theory, though you would die before reaching the singularity. However, according to a 2012 study published in the journal Nature, quantum effects would cause the event horizon to behave like a wall of fire, instantly burning you to death.

Black holes aren't fun to be around. Suction occurs when something is drawn into a vacuum, which the massive black hole is not. Rather, objects fall into them in the same way that they fall toward anything that has gravity, such as the Earth.

Cygnus X-1 is the first object to be classified as a black hole. Stephen Hawking and fellow physicist Kip Thorne had a friendly wager in 1974 about Cygnus X-1, with Hawking betting that the source was not a black hole. Hawking conceded defeat in 1990.

It's possible that miniature black holes formed right after the Big Bang. Some regions of rapidly expanding space may have been squeezed into tiny, dense black holes with masses less than the sun.

A star can be ripped apart if it passes too close to a black hole.

According to astronomers, the Milky Way contains anywhere from 10 million to 1 billion stellar black holes, each with a mass of three times that of the sun.

Black holes continue to be fantastic subject matter for science fiction novels and films. Take a look at the film "Interstellar," which heavily relied on Thorne to bring science to life. As a result of Thorne's collaboration with the film's special effects team, scientists now have a better understanding of how distant stars might appear when seen near a rapidly spinning black hole.

Chapter 7: QUANTUM MECHANICS VS. EINSTEIN

Albert Einstein's mass-energy equivalence formula E = mc2 is well-known, but his work also laid the groundwork for modern quantum mechanics.

His analysis of quantum mechanics' "spookiness" led to a slew of new applications, including quantum teleportation and quantum cryptography, but he wasn't completely convinced by the theory – and the story behind that is just as fascinating as the theory itself.

Quantum mechanics is a strange science. It implies that a particle, such as an electron, can simultaneously pass through two holes.

More famously, German physicist Erwin Schrödinger's equations demonstrated that a cat could end up in a strange quantum state where it was neither alive nor dead.

All of this did not impress Einstein. He was convinced that quantum mechanics was correct, but he was desperate to find a way to "complete" it so that it made sense.

Most quantum physicists at the time followed the "shut up and calculate" philosophy: get on with the job and don't worry about philosophical issues; just get the results.

Gaining momentum (and position)

Heisenberg's Uncertainty Principle, which states, among other things, that it is impossible to measure both the position and the momentum of a particle simultaneously to arbitrary accuracy, was used against Einstein by his opponents.

When someone measures a particle's position, the particle is disturbed, and its momentum changes. How can those two things be defined together if they can't be measured at the same time?

Opponents assumed Einstein didn't understand quantum mechanics, but he knew the issue was more serious.

Then, oh my goodness, Eureka! Einstein came up with a way to explain quantum mechanics' problems in 1935. He'd make a compelling case for how the position could be measured without disturbing the particle!

Quantum entanglement was discovered by Einstein and American physicists Boris Podolsky and Nathan Rosen. Quantum entanglement between two particles means that the quantum wave function describing them can't be mathematically factored into two parts, one for each particle. This has significant ramifications. Entanglement occurs when two particles become especially connected in a "spooky" way that was eventually revealed by Einstein's arguments and subsequent experiments.

Einstein, Podolsky, and Rosen (known as EPR) realized that quantum mechanics predicted entangled states, in which two particles' positions and momenta are perfectly correlated no matter how far apart they are.

That was important to Einstein, who believed that whatever was done to the first particle would not cause any immediate disturbance to the second particle. This was dubbed "no-spooky-action-at-a-distance" by him.

Assume that a girl named Alice measures the position of the first particle while a boy named Bob measures the position of the second particle at the same time. Because of the perfect correlation, Alice knows the result of Bob's measurement as soon as she takes her own measurement.

Her prediction for Einstein's magical entangled states is dead on – no errors at all.

Then Einstein argued that this could only happen because Bob's particle had the exact position predicted by Alice. Nothing can change at Bob's location as a result of Alice's measurement, which has no effect on the second particle. Because the measurements of Bob and Alice are separated by space, Einstein deduced that a hidden variable must exist to describe the precisely specified value of the position of the second particle measured by Bob.

Similarly, Alice can now predict Bob's particle's momentum with absolute precision without disturbing it. Then, assuming no strange behavior, Einstein claimed that, regardless of Alice's measurement, the momentum of Bob's particle could be precisely specified.

As a result, Bob's particle has precise values for both position and momentum, which violates the Heisenberg Uncertainty Principle.

Resolving spooky action

The contradiction between quantum mechanics as we know it and the assumption of "no-spooky-action-at-a-distance" was illustrated by Einstein's argument. Einstein believed that the simplest solution to the problem was to introduce hidden variables that were consistent with no spooky action, completing quantum mechanics.

Of course, the most straightforward solution is that Einstein's entanglement does not exist in nature. There have been suggestions that entanglement decays with the spatial separation of the particles, in which case quantum mechanics and spooky action would be incompatible.

Einstein's entanglement had to be confirmed experimentally.

Chien-Shiung Wu, also known as Madame Wu or the First Lady of Physics, from Columbia University was the first to demonstrate Einstein's entanglement in the laboratory. She demonstrated an Einstein-type correlation between the polarization of two well-separated photons, which are light particles that are tiny and localized.

Einstein was taken seriously by John Bell, a physicist at CERN, who wanted to develop a hidden variable theory along the lines suggested by Einstein.

He examined Madame Wu's states, but when he looked closely at their predictions for some minor adjustments in measurements, he discovered an unexpected result.

Chien-Shiung Wu.

Finding such a hidden variable theory would be impossible, according to quantum mechanics. For Einstein's hidden variables and quantum mechanics, laboratory measurements would yield different results.

This meant that either quantum mechanics was incorrect, or that any hidden variable theory capable of completing quantum mechanics had to allow for "spooky-action-at-a-distance."

Back to the lab

In a nutshell, experimentalists John Clauser, Alain Aspect, Anton Zeilinger, Paul Kwiat, and colleagues tested Einstein's hidden variable theories using the Bell proposal. So far, all evidence points to quantum mechanics. When two particles become entangled, it appears that whatever happens to one can instantly affect the other, even if the particles are separated!

Have experiments shattered Einstein's hopes for a better theory?

That's not the case. Photons, not massive particles like electrons or atoms, have been the focus of previous experiments. They also don't work with very large systems. As a result, I don't believe Einstein will give up just yet. He believes that the laws for real particles are different.

Australian scientists are experimenting with atoms and even miniature objects that have been cooled to the point where they have lost all thermal jittering to test Einstein's and Bell's theories. Who knows what they'll come across?

What about my contribution? After noticing scientists could amplify and detect the tiny quantum fluctuations of optical amplitudes while working with squeezed states of light in the 1980s, I came up with a way to test for the original Einstein's entanglement.

These are equivalent to "position" and "momentum" in quantum mechanics, and the experiment opened up a whole new way of testing Einstein's entanglement.

Experiments in a variety of environments have since confirmed this mesoscopic type of Einstein's entanglement, bringing us closer to understanding Schrödinger's cat.

Chapter 8: YOU CAN BE IN TWO PLACES AT ONCE: EINSTEIN

The scientific breakthrough of the year is a device that exists in two states at the same time and, by chance, proves that Albert Einstein was correct when he thought he was wrong. The machine, which is made up of a sliver of wafer-thin metal, is the first man-made device to be controlled by the mysterious quantum forces that govern atoms and subatomic particles.

Ordinary objects follow the laws of Newtonian physics, named after Sir Isaac Newton, but these rules break down on the subatomic scale, necessitating the creation of a new branch of theoretical physics to explain what happens at this microscopic level.

Einstein was the first to accept quantum physics, but he later rejected it because it made everything unpredictable – "God does not play dice with the universe," he famously said. However, over the last few years, a variety of effects that can only be explained by quantum mechanics have been observed, and in March, scientists were able to construct the first device that appeared to follow the quantum rules that Einstein was the first to realize applied to light waves.

The breakthrough, which was named the most important of the year by the journal Science, paves the way for a variety of practical developments, including quantum computers that are far faster than conventional processors and can never be hacked because they handle and transmit data using an unbreakable form of encryption.

"A very small thing can absorb energy only in discrete amounts, can never sit perfectly still, and can be in two places at the same time," Adrian Cho, a Science writer, explained. "Scientists have demonstrated quantum effects in the motion of a human-made object for the first time. It opens up a wide range of possibilities, from new experiments combining quantum control of light, electrical currents, and motion to, perhaps someday, tests of quantum mechanics' limits and our perception of reality."

Physicists Andrew Cleland and John Martinis of the University of California at Santa Barbara made the breakthrough. Their device was a tiny metal paddle made of semiconductor material that could be seen with the naked eye. They made the device vibrate by getting thicker and thinner at a frequency of about 6 billion times per second, producing a detectable electric current, by supercooling it to just above absolute zero (minus 273C) and then raising its energy by a "single quantum." They even got it to vibrate in two different energy states at the same time, both a lot and a little – a phenomenon only allowed by quantum mechanics.

"With a tiny object like this one, physicists still haven't achieved a two-places-at-once state," Mr Cho said. "However, now that they've achieved this most basic state of quantum motion, it appears to be a lot more achievable."

Chapter 9: SCHRÖDINGER'S CAT THEORY

Schrödinger's cat is a thought experiment in quantum mechanics that illustrates an apparent paradox of quantum superposition. As a result of being linked to a random subatomic event that may or may not occur, a hypothetical cat may be considered both alive and dead in the thought experiment.

In a discussion with Albert Einstein in 1935, Austrian-Irish physicist Erwin Schrödinger devised this thought experiment to illustrate what Schrödinger saw as the problems with the Copenhagen interpretation of quantum mechanics. The scenario is frequently mentioned in theoretical discussions of quantum mechanics interpretations, especially in situations involving the measurement problem.

E. Schrödinger proposed his famous cat thought experiment in 1935, implying but not explaining how a measurement converts the probable states of an atom into the actual state of a cat (alive or dead). Rather than using quantum mechanics (which is the most common approach), I use metrology to provide an out-of-the-box, logically consistent explanation (the science of physical measurement).

An atom (or other entity) is in a superposition, or probabilistic combination of all possible states, according to formal quantum mechanics. According to this widely held belief, every state is possible, but no state is real until it is measured. How does a measurement transform multiple possible states into a single true state? The tongue-in-cheek thought experiment of E. Schrödinger investigates this by comparing the quantum superposition of an atom to the physical measurement of a cat's state.

All of Schrödinger's information on his thought experiment is as follows: "One can even set up quite ridiculous cases." A cat is imprisoned in a steel chamber with the following diabolical device (which must be protected from the cat's direct interference): In a Geiger counter, there is a tiny amount of radioactive substance, so small that one of the atoms may decay in an hour, but equally likely, none; if this happens, the counter tube discharges, and a hammer is released through a relay, shattering a small flask of hydrocyanic acid. If this entire system has been left alone for an hour, the cat is still alive if no atom has decayed in the meantime.

It would have been poisoned by the first atomic decay. The function of the entire system would express this by mixing or smearing out equal parts of living and dead cats (pardon the expression)."

By correlating two observations (measurements): an atom's decay and a cat's death, Schrödinger's experiment appears to contrast the two distributions, the probabilistic time distribution of an atom's state and a cat's binary state (alive or dead). The actual time of each cat's death is a fixed time after the time of an atom's decay due to the apparatus (due to the "diabolical device"). Each of these equal but time-shifted distributions has a mean time of one hour, according to Schrödinger. The maximum extent of each cat's actual time of death is estimated to be two hours with a mean time of one hour.

Next, Schrödinger proposed that a human observer the cat's state for one hour. The observed state of one cat over time is a one-time observation of a third distribution. This can be compared to the cat's current condition (second distribution). This comparison is more intriguing because it appears to describe how a cat's state is measured.

All measurements, according to L. Euler, are relative: "It is impossible to determine or measure one quantity without assuming the existence of another quantity of the same type and calculating the ratio between the quantity being measured and that quantity." Calibration, which correlates the measuring apparatus intervals (in this experiment, the time between observations) to a time reference, improves the accuracy of each human observation of a cat (e.g., one second). This non-local intermediate is required to keep Euler's relative quantity ratios intact.

The third distribution of each cat's observed death time is related to both the actual death time and the frequency with which the human observes the cat (i.e., the time between human observations). The time between observations in Schrödinger's experiment is given as one hour. Then, about any possible cat's actual death time, the accuracy of this one observation is +/- one hour (about two hours).

The third distribution of the time between observations is uncorrelated with the first distribution of the atom's decay time when the proposed times (both one hour) are used. This happens because the accuracy of +/- one hour between observations is nearly as wide as the decay distribution of an atom (two hours). In its current form, this thought experiment consists of two equal but time-shifted distributions of atom decay and cat death (one function) and one unrelated human observation (a different function). However, because physical reality does not allow for the cat's state to be a superposition, this experiment has piqued interest for 85 years. What's missing here?

The first human observation of a dead cat following a life observation is timed about the start of the experiment (alive cat). These time of death observations reveal that the distribution across multiple experiments is correlated with the function of an atom's probabilities and that each cat's state is binary. The observed time of death differs from the actual time of death. Because the observed time of death is linked to the time interval between observations and the defined time reference.

Consider the experiment in which the observer (or measuring device) keeps track of how many times the cat is examined. During the approximately two-hour maximum time to complete one experiment, these observations result in a sequence of alive states followed by one dead state. Counting this series of alive observations yields a magnitude, but not a measurement that can be correlated to a physical reference, as Euler requires. The time between observations must be defined and controlled, which necessitates the calibration of the observer/measuring apparatus to a time reference. Setting and maintaining the time between observations to 10s (s = second, the time reference) is an example of calibration in Schrödinger's experiment. The maximum variation between the observed time of death and the actual time of death is then +/-10s (i.e., accuracy). Because the number of observations is unrelated to any reference, it is not a relative measurement as defined by Euler. This count becomes a sum of the counted time between observations after calibration, where each time between observations is correlated to a time reference (second).

Calibration is required to determine a cat's death time (i.e., a relative measurement). Schrödinger was correct—the atom and the state of the actual cat have the same function. However, a relative measurement, which includes calibration to a reference, is used to produce the observed cat's state. Only when the second function occurs does a relative measurement, which is physical reality, exist. Quantum mechanics cannot represent a relative measurement because it does not deal with a reference. We can only see the probabilities of a superposition without a relative measurement.

Chapter 10: THE MYSTERY AND SCIENCE BEHIND THE LAW OF ATTRACTION

According to the Law of Attraction, we attract into our lives whatever we focus on. Quantum physics teaches that nothing is set in stone, that there are no boundaries, and that anything is made up of vibrating energy. It is never important to feel trapped with an undesirable existence by realizing that everything is Energy in a state of possibility and by using the Law of Attraction to attract what we concentrate on into our lives.

We are the Universe's Creators. Newton's classical physics takes a material viewpoint, in which the Universe is made up of isolated, stable, and unchangeable building blocks. Quantum physics adopts a spiritual viewpoint in which there are no discrete parts and everything is in constant flux.

The real universe is a whirlpool of energy that bursts in and out of existence all the time. We turn this ever-changing Energy into tangible reality through our thoughts. As a result, our thoughts will shape our reality. With quantum physics, science is moving away from the idea that humans are helpless victims and toward the realization that we are completely capable creators of our lives and the world around us.

We were irrelevant cogs in Newton's Universal Machine. We are the Universe's Creators, thanks to quantum physics.

All is made up of energy. The relationship between Energy and matter is explained by Einstein's 1905 formula $E = mc2$, which states that energy and matter are interchangeable – that, in reality, everything is Energy – dancing, fluid, ever-changing Energy.

Our thoughts have an impact on this energy. It can be shaped, formed, and molded. Through our thoughts, we shape, form, and mold the energy of the Universe as Creators. The Energy of our thoughts is transformed into the Energy of our life. The Physics of Possibility is a book about the science of possibility. What the Bleep Do We Know!?, a popular film, clarifies that quantum physics is the physics of possibility. We've been socialized to believe that the external world is more real than our own. Quantum mechanics, on the other hand, suggests the exact opposite. It is said that what occurs on the inside determines what occurs on the outside. It is said that our thoughts form our universe.

Everything is possible because nothing is fixed and everything is in a state of potential. We can call anything into existence if we understand that anything is possible and focus our thoughts on what we want to attract.

My ten-year-old next-door neighbor adores the phrase "It could happen!" He probably doesn't understand the physics of possibility, but he puts it into practice with his positive attitude. He reminds me to keep an open mind. Nothing is impossible, he reminds me.

Turning Dreams into Reality Infinite potential exists in infinite abundance in the Universe. We have the power to manifest our wishes, to make our dreams a reality, by focusing our thoughts. We can be, do, and have whatever we desire by focusing our thoughts.

Chapter 11: EXAMPLES OF QUANTUM PHYSICS IN EVERYDAY LIFE

When it comes to Quantum Physics, the topic is often found to be too isolated to be addressed. We might discuss Einstein or Schrödinger. What else is there to say? The above-mentioned point may be the end of the Quantum Physics debate. Most of us would be utterly ignorant of Quantum Physics if anyone asked us about it in our everyday lives or real-life circumstances. When you hear about the real-world applications of quantum physics, you'll be shocked to learn that some of them were right in front of your eyes! Let's start by looking at some of the items that rely on Quantum Physics to work.

1. Toaster

Quantum Physics is the only reason that the bread toast you eat while sipping your morning tea makes it to your plate. To toast a slice of bread, the toaster's heating element glows red. Toasters are widely credited with being the inspiration for the invention of Quantum Mechanics. The toaster's rod heats up, toasting the bread in the process.

2. Fluorescent Light

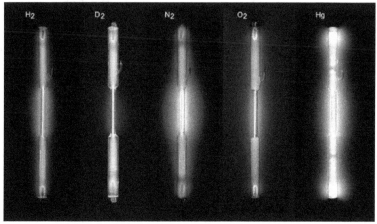

The light coming from the tubes or those curly bulbs is entirely due to a quantum phenomenon. A small amount of mercury vapor is excited into the plasma by fluorescent lighting. Mercury has the capability of emitting visible light. So, the next time you switch on your room lights at night, remember to thank Quantum Physics.

3. Computer & Mobile Phone

Quantum physics is the foundation of the entire computing universe. The wave existence of electrons is discussed in quantum physics, and this is the foundation of the band structure of solid objects on which semiconductor-based electronics are made. Not to mention the fact that we can control the electrical properties of silicon only because we can research electron wave nature. As the band composition changes, so does the conductivity. What can be done to modify the band's structure? Quantum Physics, of course, has the solution!

4. Biological Compass

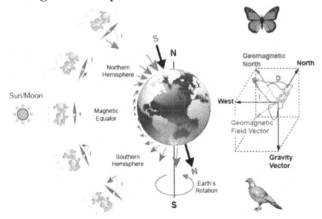

You are completely incorrect if you believe that Quantum Physics has only been applied to humans. Scientists believe that birds such as the European Robin use Quantum Physics to migrate. Cryptochrome, a light-sensitive protein, contains electrons. After passing through the bird's eyes, photons collide with cryptochrome, allowing radicals to be published. The bird can "see" a magnetic map due to these radicals. According to another hypothesis, the birds' beaks contain magnetic minerals. Magnetic compasses are used by crustaceans, lizards, insects, and even some mammals. You may be shocked to learn that the same form of cryptochrome that flies use for navigation has been discovered in the human eye! Its application, however, is uncertain.

5. Transistor

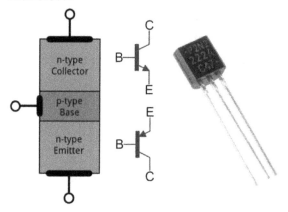

Transistors are used to amplify or alter electrical signals and power, and they have a broad variety of applications. When we study the structure of transistors closely, we can see that they are made up of layers of silicon and other components. Millions of these are used to produce computer chips, which are the brains behind many of the electronic devices that have become indispensable to human life. These chips, as well as desktops, tablets, laptops, smartphones, and other gadgets, would not have existed if Quantum Physics had not played a role in their growth.

6. Laser

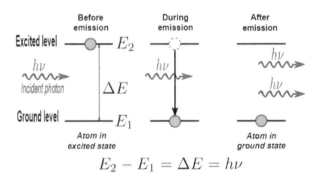

$$E_2 - E_1 = \Delta E = h\nu$$

The laser's working philosophy is focused on quantum physics. Lasers use spontaneous emission, thermal emission, and fluorescence to work. When an electron is excited, it jumps to a high-energy state. It will, however, not remain in the high-energy state for long, and will instead return to the lower-energy state, which is more stable and therefore emit light. External photons with a frequency correlated with the atomic transition often influence the quantum mechanical state of the atom.

7. Microscopy

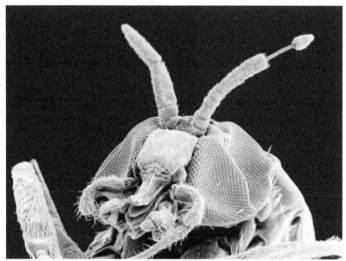

The fundamental principles of Quantum Physics have enhanced electron microscopy. The imaging of biological samples has improved thanks to quantum physics and electron microscopy. Furthermore, in differential interference contrast microscopy, the beam of photons produces an interference pattern, which is then analyzed. All-in-all, quantum physics has greatly advanced microscopy, allowing for the extraction of a vast amount of information from a sample.

8. Global Positioning System (GPS)

With the aid of Quantum Physics, navigating to unknown places has never been simpler. When using a smartphone for navigation, the phone's GPS receiver is in charge of picking up signals from various clocks. Different arrival times from different satellites are used to measure the distance and time between your current position and the destination. Furthermore, each satellite's distance from your current position is determined. Every satellite has an atomic clock that is based solely on quantum physics.

9. Magnetic Resonance Imaging

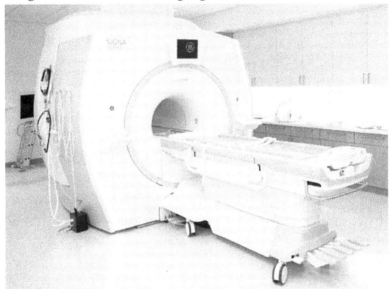

The reversal of electron spins in hydrogen nuclei is used in Magnetic Resonance Imaging, also known as Nuclear Magnetic Resonance. So, essentially, we're talking about energy shifts, which is just one of Quantum Physics' applications. The use of MRI can easily be used to investigate soft tissues. The diagnosis and treatment of certain life-threatening illnesses have been made possible thanks to quantum physics.

10. Telecommunication

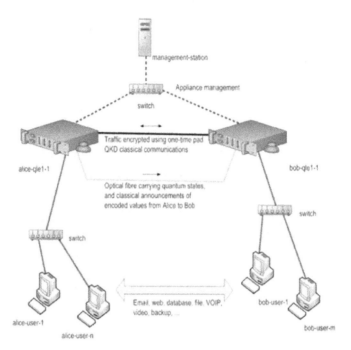

Because of the crucial role of quantum physics in communication, it has become extremely simple. Two-way and fast communication is now possible thanks to fiber optic telecommunication. Only lasers, which are quantum physics instruments, make fiber optic telecommunication possible.

CONCLUSION

Before I go, I'd like to bring up another fascinating field of quantum theory speculation. It has been proposed that if the entire universe (space, time, energy, and everything else) is quantized and made up of indivisible fundamental particles, it has a finite number of components and states, similar to how a computer program has bits and pixels. In theory, this renders the universe "computable," leading others to speculate that all of life as we know it may be part of a massive Matrix-like computer simulation, with individual parts taking on definite form only when observed. Perhaps just speculation, but it's an interesting one, and one that's becoming increasingly difficult to refute as the nuances of quantum theory are ironed out.

CPSIA information can be obtained
at www.ICGtesting.com
Printed in the USA
LVHW080533280521
688664LV00007B/774

9 781802 101485